AF151160

BEI GRIN MACHT SICH IHR WISSEN BEZAHLT

- Wir veröffentlichen Ihre Hausarbeit,
  Bachelor- und Masterarbeit

- Ihr eigenes eBook und Buch -
  weltweit in allen wichtigen Shops

- Verdienen Sie an jedem Verkauf

Jetzt bei www.GRIN.com hochladen und kostenlos publizieren

Suleiman Usman

Bio-organic Manure: A Manual Guide

GRIN Verlag

Bibliografische Information der Deutschen Nationalbibliothek:

Die Deutsche Bibliothek verzeichnet diese Publikation in der Deutschen National-
bibliografie; detaillierte bibliografische Daten sind im Internet über http://dnb.d-
nb.de/ abrufbar.

Impressum:

Copyright © 2012 GRIN Verlag, Open Publishing GmbH
Druck und Bindung: Books on Demand GmbH, Norderstedt Germany
ISBN: 978-3-656-11447-5

BIO-ORGANIC MANURE

MANUAL GUIDE

2011

Prepared

By

Suleiman Usman

PhD Student
Natural Resource Institute
The University of Greenwich UK

MANUAL GUIDE INFORMATION

SOME OF THE BIO-ORGANIC MANURES
PRODUCED FOR MARKETING PURPOSES

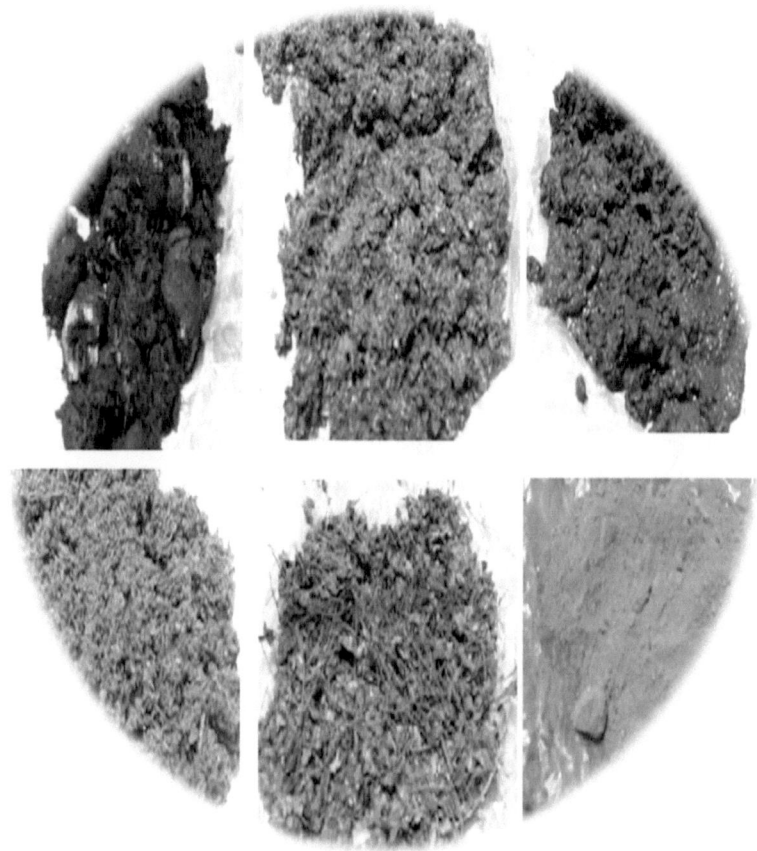

SOME OF THE BIO-ORGANIC MANURE
PRODUCED UNDER LOCAL PRODUCTION

BIO-ORGANIC MANURE FROM ANIMAL AND PLANT MATERIALS

INTRODUCTION

It has been widely accepted that organic materials (plant and animal sources of organic matter) from plants and animals play important role in sustaining and improving soil structure, soil quality, soil function, soil health, soil fertility, and overall crop performance in agricultural production. Organic materials are potential important sources of micro and macro nutrients in agricultural soils environment (Hood, 2001). They affect physical, biological, chemical, and ecological processes in soil. They improve soil structural quality, soil water holding capacity, soil infiltration, soil organism biodiversity, and soil nutrient availability (FAO, 2005). The utilization of the animal and plant materials is not only an economic benefit for soil and crops but also for human being and his environment.

Bio-organic Manure (BOM) is processed from some organic materials (animal or plant or combinations of all). Process of decomposition of homogeneous and heterogeneous organic materials will be rapidly undergoing some changes due to microbial functions, temperature control, and water actions. This is different from fresh organic fertilizer where natural decay process is brought about by the action of heterogeneous microbes present in the organic matter.

Because the animals can only absorb 25% nutrients of the plant and animal feed they eat, 75% of the nutrients is discharged with animal manure which therefore contains many elements like nitrogen, phosphorus, potassium, calcium, zinc, manganese, amino acid, proteins and many others (FAO, 2005). Therefore, to produce quality organic manure, combinations of animal, plant, crop residues, wood, and other organic materials are required. The most available and affordable among the others in some part of Africa are plant materials such as *Acacia nilotica, Acacia albida, Azadracta indica*; animal materials such as cow dung, goat dung, sheep dung, donkey dung, horse dung, camel dung, poultry dung; crop materials such as rice husks, millet husks, maize husks, sorghum husks, cowpea residues, g/nut husks; wood materials such as wood ash, wood husks; house refuses; and many others.

These materials can be identifying by adapting the field and laboratory test. However, whatever materials are selected, there must be a mixture ration of both animal and plant materials for quality product to be produce.

Generally, the BOM machine produced products that are fresh manures. These manures are raw material which does not contain any chemical composition. The available pure natural nutrients are nitrogen, phosphorus, potassium, organic matter, amino acids, proteins and other ingredients. The manure machine not only creates economic benefits for the enterprise, but also of mankind made a great contribution to environmental control and protection. The result was environmental safety and friendly, no any risk impact, significant soil quality and health conditions, high crop yield and permanent soil fertility and environmental management.

LISTS OF CATEGORIES OF RECOMMENDED ORGANIC MATERIALS

1. Raw materials and their sources. The amount depends on the desired quantity of:
    a. Animals dung – cow, sheep, goat, donkey, camel, horse, poultry
    b. Plant residues – leaves, branches, and roots
    c. Crops residues – husks of millet, rice, g/nut, sorghum, cowpea
    d. Wood materials – wood ash, wood husks
    e. House refuses – any available combinations: unwanted food, leaves, fruits, etc

LISTS OF MATERIALS TOOLS

1. Tools and equipment to be use
    a. Rake
    b. Shovel: metal and woody
    c. Metal spoon: long size one
    d. Hand gloves, boots, and nose + mouth covers
    e. Sprinkler + tank of water
    f. Plastic rubber
    g. Wheelbarrow
    h. Weighing scale
    i. Bags
    j. Human supports

GENERAL PROCEDURES

The procedures described below followed the same steps as given by Department of Agriculture Philippines Coconut Authority (DAPCA, 2010): Research, Development and Extension, Agronomy and Soils Division, Training Centre.

1. Site selection: The ideal bio-organic manure environmental site should be clean, shaded, well drained and protected from all other air and insect born diseases. Water should be close to the site and regular inspection will be timed by the producer. However, an open area could be used but fencing is important to protect the inner site of the production area. Depending on the quantity and size of the site, as much as many tones can be process at a time.

2. Preparation of raw materials: Required amount of organic materials of various types should be collected as samples. These will be then mixed together in an experimental plastic reservoir or site place. Volume of water will be added regularly to enhance the functional activities of microorganisms and gaseous exchange during the mineralization and humifications of fresh organic materials.

3. Piling of materials: the following steps are required:

    a. *Step 1.* Spread 4/10 of required plant materials as the first layer followed by 2/10 of animal materials, and then 4/10 of different kinds of plant and animal mixtures. Water should be adding equivalent to 70%-80% moisture content. After a period of 10 hours, press the sample in your hand and when the water does not fall freely, then it is almost within the right moisture content. Again, apply the required amount of water hircine (depending on the amount of materials used) to mix the materials thoroughly.

    b. *Step 2.* Spread evenly any of the animal manures (cow dung, goat dung, donkey dung etc) on top and then add water to moist it again.

    c. *Step 3.* Repeat steps 1 and 2 but use any house refuse materials instead of animal manure. Apply the rice husk or millet husk on top of the animal manure.

d. **Step 4**. As the topmost layer, spread evenly wood ash and moist it.

e. **Step 5**. Cover the materials with plastic sack to conserve moisture and prevent rainwater from getting into the material.

f. **Step 6.** Maintain regular inspections and observations including moisture and temperature control consecutively up to 15 days.

g. **Step 7.** Mixing and turning over should also be maintain after 2-3 days. During these exercises, water should be added if needed and always after turning and mixing, return the plastic cover. These should be continuing for up to 5-8 weeks.

h. **Step 8.** Harvesting and processing of materials should follow after the materials are well decomposed and mixed together properly. The indications of these decomposition and mixing are dark, dark brown to black colors.

i. **Step 9.** The processed material must be store for one week under room shade.

j. **Step 10.** Finally, after one week under shade, the materials could be put into machine to produce BOM into different shapes, colors and sizes. Put BOM in 50 kg or 70 kg or 100 kg plastic sack and then seal ready for use.

QUALITY CONTROL MEASURES

For good quality bioorganic manure, the following must be observed:

1. Separate the dry materials from wet ones in the first step.

2. Monitor periodically the temperature of the site. It must be within the right temperature range ($40^{\circ}C$). Within the first week, desired temperature is $40\text{-}50^{\circ}C$. Room temperature must be maintained. This room temperature is apply to all other bioorganic production's products such as urea, ammonium bicarbonate, chloride, ammonium phosphate, potassium chloride and other materials for the multi-element compound fertilizer.

3. For a product to be registered with the national or international bodies as pure organic manure, it must have 5-10% total NPK with at least 1.5-2.0% N and at least 10-30% carbon or calcium or magnesium.

HOW MUCH BOM TO APPLY IN THE CROP FIELD

The rate of application depends on the kind of soil and crop condition to be improved. However the following should be considered whenever it comes to application in the field:

1. BOM application ranges from 10-60 (small size) bags/ha for cereals crops farms (e.g. millet farm, maize farm), 1–2 bags for irrigation system/bed, and or 1-15 kg/tree for economics trees plantation like mango, guava, orange, and cashew.

2. Initially for degraded soil, annual supply of up to 10–20 bags is important. After this application the farm will maintain its fertility for next 2–3 years without addition of any bag of BOM. However, it is also equally important to add at least half of recommended amount after 2 years to maintain permanent soil quality and fertility for required yield performances.

BIO-ORGANIC MANURE PRODUCTION MACHINE

The descriptions of functional parts of the organic production machine are as follows:

1. **Structure:** the machine production line mainly includes crusher, disc granulator, rotary dryer, rotary cooler, screening machine, vertical mixer, belt conveyor, packaging machines and fertilizer coating equipment.

2. **Pipe line**: the pipe line of the machine is used to dry the ordure from pig and chicken which have no any chemical content because the pig and chicken only can digest 75% and 25% is included in ordure so that the fertilizer from them possess of plentiful Nitrogen Phosphorus, Calcium, Organic Matter, Aminophenol, Protein and so on.

3. **Materials in machine line component:** Organic manure production line use animal dung, poultry dung, plant materials as its raw material which contains no chemical elements.

DIFFERENT TYPES OF BOM PRODUCTION MACHINES

1. **Multifunction organic (BIO) ball granulator completion machine** or – Organic fertilizer granulator or – Multifunctional Organic Manure (bio-) Ball Granulating Machine, Circle die type pellet machine: the organic fertilizer granulator is good quality with high capacity. The machine is used to make the fertilizer pellet, etc.

2. **Bioorganic fertilizer dryer machine:** the machine is used to dry the ordure from all animal and poultry dungs which have no any chemical content because animal and poultry only digest 75% nutrients and 25% is included in ordure so that the manure from them possess of plentiful nitrogen, phosphorus, calcium, organic matter, aminophenol, protein and many other.

3. **LP800 fertilizer crushing machine, Fertilizer crusher:** LP Series Crushers chain is used for the production of massive compound of broken objects. The aircraft of the machine have been in the crushing and the use of synchronous speed of high-strength wear-resistant carbide chain board, entrance and exit into and out of rational design of materials evenly broken and difficult to dip the wall, easy to clean up.

4. **Livestock dung processors:** this is screw extrusion solid-liquid separation machine which are characteristically small, low speed, simple operation, installation, maintenance convenience, the cost of the province, high efficiency, quick recovery of investment, without adding any flocculants. It would shake the original livestock water separation for solid organic manure and liquid organic manure. Liquid organic manure can be used for crops directly, the use of absorption. Solid organic fertilizer can be shipped to less fertilizer area, improved soil structure; meanwhile it can be made organic manure through fermentation.

5. **Animal manure dewater machine, Animal waste dewater machine:** The machine use the pump to pull the fowl manure, cow dung, donkey dung, sheep dung, goat dung

etc. into dewater equipment and the pure material through the screen and then press by the screw. The rotary speed of the machine can reach 45r/min, within the press of the screen and high rotary speed. The material will be dewatered by the machine, and the water will enter into the pool through screen.

6. **Fertilizer Mixing Machine/mixer, Fertilizer Mixing Machine:** Fertilizer Mixing Machine has been developed recently and adapts to over two kinds of organic manure and additive fertilizer. It can mix organic manures fully and uses novelty rotor configuration that can adjust the gap to nearly zero between rotor and shell. The machine can pulverize heave matter and use conveniently.

CROP-SOIL BENEFITS FROM BOM PRODUCTION

1. Improves quality and health of crop's seeds
2. Increases yield performance
3. Seed free from pest and diseases damages
4. Supply and maintain minerals, vitamins, and proteins compounds in seeds
5. Improves soil structure, colour and texture
6. Provides better soil aeration and increases water-holding capacity
7. Improves the functional development of soil chemical properties
8. Help to corrects and maintain proper level of acidity, alkalinity, and soil pH
9. Help to restores regular soil fertility functions by enriching soil with trace and other micronutrients and organic matter in soil zones
10. Help to restores the activities of soil biota: micro and macro organisms and their microbial balance in soil

GENERAL ADVANTAGES OF BOM PRODUCTION

1. Reduced dependence on inorganic fertilizers
2. Soil fertility management
3. Soil health, soil function, and soil quality performances
4. Soil resilience's against impact of raindrops
5. Improve soil structure, soil colour, soil texture, soil consistence and water infiltration
6. Renewable and locally available raw materials

7. Contribute to proper waste disposal

8. Generate employment opportunity

9. Contribute to biodiversity

10. Lessen environmental pollution

11. Reduce chemical pesticide, insecticide, fungicide etc. environmental hazards

12. Provide a solution to human diseases such as ulcer, cancer,

13. Improve and increase growth and crop yield performances

14. Good source of income and revenue to government

15. Environmental friendly

16. Help to preserve and sustain natural habitat

17. Help to develop and establish forest regenerations of economic trees

18. Help to reduce the impact of soil erosion and desertification

19. Poverty reductions

20. Sustain wide range of soil processes (physical, chemical, biological, and ecological)

REFERENCES USED

1. DAPCA (2010) Bio-organic Fertilizer from Coir Dust and Animal Manures. Available at: www.pca.da.gov.ph/pdf/techno/bioorg.pdf Department of Agriculture Philippines Coconut Authority (DAPCA): Research, Development and Extension, Agronomy and Soils Division, Training Centre, Davao City.

2. FAO (2005) The importance of soil organic matter: key to drought-resistant soil and sustained food production. *FAO Soil Bulletin.* Natural Resources Management and Environment Department Alexandra Bot/Jose Benites, FAO. 94 pp

3. Focus Technology Co., Ltd. China. Available at: www.made-in-china.com/.../China-Bio-Organic-Fertilizer-Biological-Manure.html

4. Hood, R. (2001) Evaluation of a new approach to the nitrogen 15Isotope dilution technique to estimate crop N uptake from organic residues in the field. *Biofertile Soils, 34:156 – 161.*

5. Pincusby, S. (2010) Organic in our midst – Myth or Possibility? *Park and Recreation Magazine March-April 2010,* Illinois, webXtra.

6. Vanashree Agriculture Pvt. Ltd. (Vanashree Agrotech) J-116, Megacentre, Magarpatta, Solapur Road, Hadapsar Pune, Maharashtra - 411 028, India Available at: www.vanashreeagrotech.com/bio-organic-fertilizers.html